LIVING IN SPACE

LIVING IN SPACE

Don Berliner

Lerner Publications Company
Minneapolis

Copyright © 1993 by Lerner Publications Company

All rights reserved. International copyright secured. No part of this book may be reproduced, stored in a retrieval system, or transmitted in any form or by any means, electronic or mechanical, including photocopying, recording, or otherwise, without the prior written permission of Lerner Publications Company, except for the inclusion of brief quotations in an acknowledged review.

Library of Congress Cataloging-in-Publication Data

Berliner, Don.
  Living in Space / Don Berliner.
    p. cm.
  Summary: Discusses such considerations for future manned spacecraft as design features, preparing and eating food, personal hygiene, interpersonal relationships, exercise, and safety.
  ISBN 0-8225-1599-7
  1. Space stations—Juvenile literature. [1. Space stations. 2. Manned space flight.] I. Title.
TL797.B47   1993
629.47'7—dc20   92-24847
                                                          CIP
                                                          AC

Manufactured in the United States of America

1 2 3 4 5 6 98 97 96 95 94 93

CONTENTS

	Introduction	7
1	Food and Clothing	15
2	Hygiene	27
3	Staying Healthy	31
4	Getting Along Together	39
5	Designing a Livable Spacecraft	47
6	Safety and Maintenance	55
	Summary	*60*
	Resources to Contact	*61*
	Index	*62*

A photo of Gemini 7 *taken from* Gemini 6 *during a rendezvous maneuver*

INTRODUCTION

When astronauts orbit the Earth, they travel at 18,000 miles (28,800 kilometers) per hour—much faster than anyone can travel in an airplane. The surprising thing is that they have no sense of speed. Astronauts feel as though they're floating, but actually they're falling. But anything falling at such a high speed follows an arc that has the same curvature as the Earth. While in orbit, the spacecraft has no weight. It remains at the same altitude as it "falls" around the earth for days, weeks, or even months. The air inside the spacecraft, however, resembles the air in an airliner—drier and slightly lower in pressure than air at sea level.

Several different types of spacecraft have been launched into orbit or flown to the Moon since the early days of space travel, and the astronauts' living conditions during flights have changed dramatically. In 1961, when cosmonaut Yuri Gagarin became the first human to orbit (circle) the Earth in *Vostok 1*, his personal comfort was of little concern. All that mattered about the huge, expensive, dangerous operation was that it be a successful mission. The goal was to launch cosmonaut (a Soviet astronaut) Gagarin into space to orbit the Earth and then return alive and well. No one worried about whether he felt cold or hot, cramped or uncomfortable, bored or hungry. Gagarin was a pioneer. He willingly traded personal comfort for the opportunity to do something historic, something he considered truly important. After all, he was in space for just 110 minutes. The veteran pilot could endure a lot of discomfort during that short time.

The same was true 10 months later when U.S. astronaut John Glenn orbited the Earth three times in a Mercury capsule called *Freedom 7*. He had known for a long time that he would be squeezed into a tiny cabin with barely enough room for him to stretch his legs. He also knew he wouldn't have anything to eat. It didn't matter to Glenn, because he went into space to work and to learn. His five-hour mission was so crowded with important tasks that nothing else mattered. He was an explorer. The important journeys of explorers and pioneers have always been filled with danger and discomfort. That's one of the reasons that astronauts and cosmonauts have been welcomed like heroes when they return from space.

Orbital flights of a few hours were strictly business. Those first crewed flights were in spacecraft that seem very primitive by current standards. In the 1960s, astronauts, engineers, and technicians were just beginning to learn the most basic things about space travel. Much had been learned on the ground through the use of simulators—devices that almost reproduce the actual conditions of a space flight. But many things could be learned only in space.

Scientists and engineers on Earth learned a great deal by monitoring the crew in space. That information helped them improve the next generation of orbital craft so that future crews could fly longer missions in greater comfort. As launch vehicles became more and more powerful, newer and better spacecraft could be boosted into Earth's orbit and beyond. The new spacecraft were bigger, heavier, and much more complicated vehicles. Larger crews could accomplish more during longer flights. A crew's workload usually includes conducting scientific experiments, performing maintenance on the spacecraft and its equipment, communicating with mission control on Earth, and navigating the craft.

But progress also brought new problems. For example, when two astronauts are confined in a spacecraft that is barely large enough for them to work in, tension can develop. What might be a minor disagreement on a short flight can become a serious problem on a longer flight. Making living conditions

Introduction 9

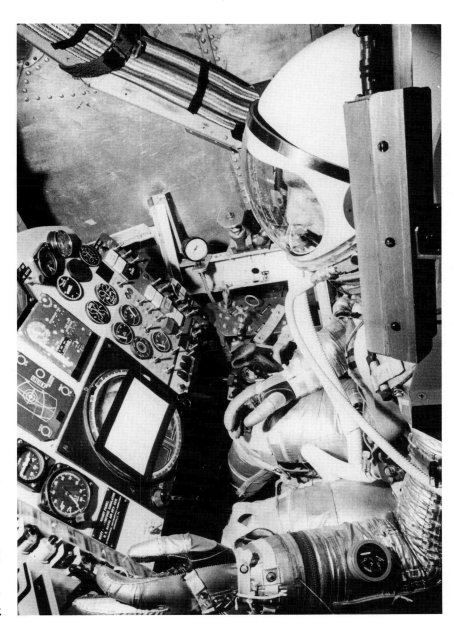

Astronaut Scott Carpenter shows the cramped quarters of early spacecraft. He is in a 1961 Mercury *spacecraft simulator.*

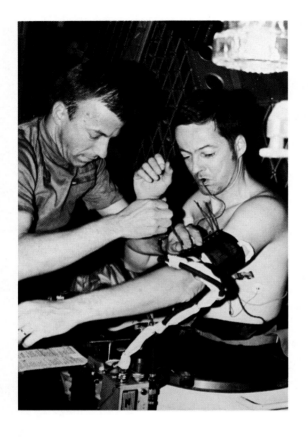

Astronaut Paul J. Weitz, left, *helps scientist-astronaut Joseph P. Kerwin,* right, *adjust a blood-pressure cuff during a lower body negative pressure experiment aboard the* Skylab *space station. Kerwin is lying in the lower body pressure device. The purpose of the experiment is to provide in-flight data on blood pressure, heart rate, body temperature, leg volume changes, and body weight.*

as normal as possible can ease some of the tension. As space flights grow longer and involve more astronauts, attention to comfort becomes more and more important.

Apollo 7, which was boosted into orbit in 1968, was the first U.S. spacecraft in which astronauts could move around in the command module, the area in which the crew spends most of its time. But the inside of *Apollo* was filled with technical equipment that was not especially pleasant to look at. Designers had put most of their effort into creating an efficient machine, and they paid little attention to crew comfort or a pleasing appearance. Since its longest flight—to the Moon and back—

Skylab *space station, as seen from the command module. The protrusions are huge solar panels attached to the space station.*

Shannon W. Lucid floats aboard a KC-135 aircraft during zero gravity training.

would last just 12 days, NASA (National Aeronautics and Space Administration, a U.S. government agency) felt no need to make the craft anything but functional.

Skylab, America's first space station, was launched in 1973. In terms of comfort, *Skylab* was a big step forward. Weighing 44,000 pounds (20,000 kilograms), it had plenty of room for its three-man crew to work and to relax.

Those involved in space research learned a great deal from studying *Skylab* and the larger, more complicated craft that followed. Both NASA and the former Soviet space agency strove to make their future crewed spacecraft better places for men and women to spend long periods of time.

Perhaps the strangest and most interesting feature about living in space is that people are weightless because there is no gravity. This condition, which is called zero gravity, makes it possible for astronauts to fly through the cabin simply by pushing off from a wall or chair. It's almost like swimming underwater, except that there is no up or down.

Although it might sound like weightlessness is great fun, the condition also has its drawbacks. Being weightless can make astronauts feel congested, as though they have colds. Also, an astronaut's face might become puffy. One former Soviet cosmonaut said, "Our weightlessness isn't that much of a pleasure. Our faces have begun to swell, so much that looking in the mirror, I fail to recognize myself. I keep bumping into things, mostly with my head."

Others report short-term hallucinations, loss of appetite, vomiting, vision problems (loss of depth perception and eye fatigue), and changes in reaction speed. These problems are more serious for some people than for others. Most symptoms last a few days and then disappear. In some cases, however, it may be necessary for a crew member to wear a special suit that maintains pressure on certain parts of the body. For example, the force of gravity that builds up during reentry into the atmosphere can cause blood to flow away from the head and toward the legs. This can cause an astronaut to black out. To avoid that situation, the legs of a flight suit can be inflated. The pressure from the inflated pants legs prevents blackouts from happening.

After the eight-month *Salyut 7* mission, cosmonaut Oleg Yur'yevich Atkov said, "Our adaptation to weightlessness was surprisingly easy and fast. Virtually by the second day aboard the station, we no longer felt the painful throbbing in our heads from which many Soviet cosmonauts and American astronauts have previously suffered. The special training sessions we underwent at the Cosmonaut Training Center probably better conditioned us for this extended mission."

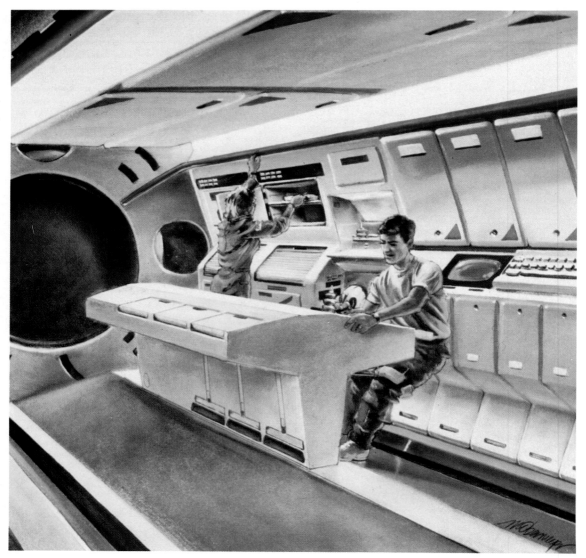

This is an artist's rendering of the galley, or kitchen, for a permanently crewed space station. At right, an astronaut, who is held in place by restraints, sits at a workstation while another crew member prepares a meal.

1
FOOD AND CLOTHING

In space, as on Earth, eating is an important part of life. Every living creature, from a single-celled amoeba to a 325-pound (145-kg) football player, eats several times a day from birth to death. Not only is eating absolutely vital for survival, it is also enjoyable, and it can even affect morale. If you don't eat enough food, or if you don't like the food you eat, your mental and emotional well-being can be affected. When that happens, you won't be able to work efficiently.

Eating in Space

During the early space flights in the 1960s, crews did not complain about their food, even though it didn't taste very good. Designed for practicality, food on spacecraft was compressed, processed, and packaged to take up a minimum amount of space and to last for a long time without spoiling. The only problem, as one official said, was "to influence the crews to eat [it]!" In preparation for the first flight to the Moon, a NASA official—known as a man who would eat almost anything—ate only the food planned for the Apollo mission for four days. He found it *very* unpleasant, and his experience led to more concern for the crew's needs.

Dietitians and experienced astronauts now test the types, amounts, and varieties of food for each astronaut. They test the food both on Earth and in space to see if it provides energy and nutrients and satisfies the personal needs of crew members.

The larger size of current spacecraft has made meal planning much easier than it was in the early days of space flight. The tiny capsules of the Mercury and Gemini programs had little space to store bulky food and a limited ability to sustain more than the weight of the craft and its crew in orbit. As a result, the astronauts were stuck with food and drinks that they found rather disagreeable. Since flights were fairly short, however, the crews put up with the discomfort and made plans to gorge themselves with "real" food when they returned home.

On one *Gemini* two-man flight, a carefully planned experiment was ruined when one astronaut playfully asked the other for his favorite sandwich. He was promptly handed a smuggled corned-beef-on-rye. Scientists on the ground were furious to see their experiment, which dealt with the effects of a special diet, wrecked—but the astronauts were amused.

Types of Food

Meals on the most recent crewed flights have included food the astronauts would ordinarily eat on Earth: meats, dairy products, baked goods, fruits, vegetables, eggs, soups, beverages, desserts, and snacks. Dietitians now try to make space food resemble Earth food as closely as possible. Unusual tastes and textures tend to make the astronauts feel less "at home" in space. Meals are even scheduled much the same as they are on Earth—the crew has breakfast, lunch, and dinner—even though sunrise and sunset are only 45 minutes apart in orbit.

As space flights stretch into weeks and months (the longest flight, by a Soviet *Soyuz*, lasted 366 days), astronauts tend to tire of some foods, crave other foods, lose their appetites, or want foods that are spicier than usual. Mission planners have learned that they must work out compromises between food value and crew morale.

Amounts of Food

While the particular foods each astronaut eats vary from day to day, the total food value is carefully monitored and controlled to make sure that each crew member has a healthy diet. Soviet experience has shown that astronauts, or cosmonauts, need about

3,000 calories per day—about 140 grams of protein, 100 grams of fat, and 400 grams of carbohydrates. These amounts differ from recommended diets on Earth because the workload in space is different from that on Earth.

An astronaut should consume about 1.5 pounds (700 g) of food in one day. Conditions in space also require that crews drink 2.5 quarts (2.4 liters) of water each day, which is much more than most people usually drink in a day.

Individual Preferences

Each astronaut is an individual. The longer the flight, the more important individual differences can become, and the more important it is to deal with them.

In April 1990, space shuttle *Discovery* carried the Hubble Space Telescope and launched it into orbit. The five-member crew on mission STS-31 (Space Transportation System flight number 31) expressed a wide variety of food preferences. For example, here is what the crew aboard *Discovery* ate for breakfast on the fourth day of the mission:

Commander Loren Shriver: sausage patty, Mexican scrambled eggs, blueberry yogurt, apple cider, and decaffeinated coffee with sugar.

Pilot Charles Bolden: chicken consommé, dried apricots, beef patty, seasoned scrambled eggs, blueberry yogurt, and grapefruit drink.

Mission Specialist Bruce McCandless: trail mix, beef patty, oatmeal with brown sugar, orange juice, and plain tea.

Mission Specialist Steven Hawley: sausage patty, scrambled eggs, grapefruit drink, and chocolate instant breakfast.

Mission Specialist Kathryn Sullivan: pineapple, oatmeal with raisins, orange juice, and black coffee.

None of these breakfasts may be what you would have chosen to eat, and some of the items—such as chicken consommé—may seem strange for anyone's breakfast. But remember that the meals were designed to fit each astronaut's own tastes, and breakfast may have been eaten shortly after sunset.

The experience of both American and Soviet crews and scientists suggests that it is okay to repeat meals every five or six days without creating great dissatisfaction among the crew.

Seven-Day Diet

To show how one astronaut's diet can vary from day to day, here is what Mission Specialist Kathryn Sullivan had for dinner during the week-long flight:

Days 1 and 6: chicken teriyaki, rice and chicken, asparagus, and black coffee.

Days 2 and 7: meatballs with barbecue sauce, potatoes au gratin, green beans with mushrooms, and black coffee.

Day 3: sweet and sour chicken, green beans, broccoli, strawberries, and black coffee.

Day 4: beef tips with mushrooms, rice pilaf, Italian vegetables, and black coffee.

Day 5: beef goulash, creamed spinach, and black coffee.

While Americans and the former Soviets have generally agreed on how to plan meals, their menus do reveal some cultural differences. For example, here are two breakfast menus for cosmonauts: (1) chicken with prunes, bread, candy, and coffee with milk; (2) pork with sweet pepper, Russian cheese, honeycake, prunes, and coffee. Cosmonauts have more and smaller meals than American astronauts do. Typical Soviet dinners might include sauerkraut soup, roast beef with mashed potatoes, bread, prunes with nuts, and candied fruit; or ham, borscht (thick soup) with smoked foods, beef with mashed potatoes, rye bread, cookies with cheese, and apple juice. A cosmonaut's diet is much more of the "meat and potatoes" kind than is an astronaut's. All travelers in space get vitamins and food supplements, but only cosmonauts get a small amount of vodka and brandy to add to their diets.

Space shuttles are equipped with pantries where food, including snacks, is stored. Astronauts enjoy snacks and eat them whenever they feel a bit hungry, or sometimes when they're a little bored. On STS-31, the array of snacks included almonds, dried apricots, brownies, butter cookies, candy-coated chocolates and peanuts, cashew nuts, chocolate-covered cookies, crackers, granola bars, macadamia nuts, shortbread cookies, and trail mix.

Food and Clothing 19

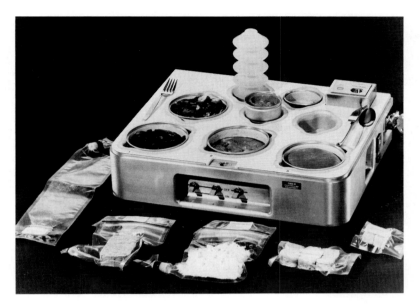

The astronauts at right are taste-testing food that may be used on long-term missions. Varying degrees of enthusiasm are evident. A close-up view of a food tray, utensils, and extra packets of space food are seen above.

Hal Davis, a lab technician with a space botany program, tends lettuce that is being grown in a hydroponic cabinet, where plants grow in nutrient solutions. The program is designed to develop techniques for growing vegetables and grain in space. The goal is to create "space gardens" so astronauts can have fresh food during long missions.

Of course, no astronaut is allowed to ruin the plans and experiments of the dietitians by stuffing himself or herself with snacks. However, there must be a great temptation to just sit by a window and gaze down at the continuously changing view of Earth while eating chocolate-covered cookies. Fortunately, the pantry also stocks things like apples and bananas.

If current plans become reality, the American space station *Freedom*, scheduled to be operational in the mid-1990s, will be able to provide ethnic foods for astronauts from other countries, and even a cake for a crew member who is celebrating a birthday.

Food Preparation

The first crewed space flights relied on foods and drinks that were either ready to eat as they came out of their packages or that could be prepared by just adding water and shaking to mix the ingredients. Although such food was certainly convenient and took up little of the crew's time, it provided little or no enjoyment. One early astronaut told the author, "I'll never drink that stuff again as long as I live!"

The new, larger spacecraft are roomier inside and have enough electrical power to run various types of cooking equipment, such as microwave ovens. Due to these improvements, the "kitchen/dining room" has developed into a useful place almost like home.

Crew members can eat their meals alone or together as a group. Usually, each person prepares his or her own food. Since storage space is limited, dietitians select foods that produce little waste. Any leftover waste is carefully wrapped to be returned to Earth. In the future, trash compactors will probably be installed in spacecraft so that waste products will take up even less room.

When completed, space station *Freedom* will provide a permanent structure in space for scientific research. Space shuttles will bring up replacement crews, supplies, and new equipment.

Growing Food aboard a Spacecraft

As space missions become longer, more storage space will be needed to store food for the crew. If most of the food needed by the crew could be grown on the spacecraft, then less food would have to be carried aloft at the start of a mission. Experiments for growing food aboard a spacecraft have been underway for several years. Complete success remains far in the future, but progress is being made.

Growing plants aboard a spacecraft requires controlled temperature, soil nutrients, light, and moisture. The conditions that suit plants, however, may not be healthy or comfortable for the crew. Special containers, perhaps resembling a greenhouse, may be needed for growing plants in space. Space farmers also face other problems, such as the inability of water to run downward. There is no "down" in zero gravity. Because of the absence of gravity, the ability of some plants to process nutrients may slow down or even stop. The test use of a small centrifuge (a machine that can simulate the effects of gravity) to spin experimental plants and create artificial gravity has shown that the roots from some seeds grow twice as fast under this treatment than they do in zero gravity.

On their seven- and eight-month *Salyut* space station missions, the former Soviets tried to grow a wide variety

Tom Slavin, a life sciences design engineer, demonstrates how an astronaut might conduct plant-growth experiments aboard space station Freedom.

of plants, including fruits (strawberries and tomatoes), vegetables (onions, cucumbers, radishes, carrots, peas, lettuce, parsley, peppers, and kale), grains (wheat and oats), and herbs. Some of these plants grew well and were eaten aboard *Salyut*, but others grew poorly, and still others died. Much more research must be done before a spacecraft can become a self-sustaining unit, but scientists feel they are headed in the right direction.

NASA, for example, is working on a Controlled Ecological Life Support System (CELSS). The project involves growing crops aboard a space station, converting them into food for the crew, and using the inedible materials for fertilizer. A test program has shown encouraging results. If such a plan can be developed, a future space station could supply many of its crew's food needs.

Clothing

Just as food is important during a long space mission, so is comfortable, functional clothing. On Earth most people are free to choose the clothes they wear. Factors such as style, cost, comfort, durability, and practicality usually influence clothing decisions. For members of a space crew, however, clothing decisions are based primarily on whether the astronauts will be able to work effectively. Engineers, not fashion designers, decide what astronauts wear in space.

The first space crews wore heavy, bulky suits before and during their flights. They could not change clothes until they returned to Earth. As conditions in the cabin were made more like

Two members of a space shuttle crew sit at their posts during reentry into the Earth's atmosphere.

Owen Garriott deploys a particle collection experiment during a space walk on Skylab, *above. Astronaut James Wetherbee, right, plays the "drums" in the copilot's station aboard space shuttle STS-31.*

those on Earth, however, the crew could remove their space suits and work in a "shirtsleeve" environment once they got into orbit. This made working in space easier, less tiring, and more like working in ordinary life.

In contrast to the comfortable clothing worn in the cabin, however, astronauts need specially designed clothing for extravehicular activity (space walks) because the outside pressure is zero. A space suit is designed to withstand the lack of outside air pressure. It contains a life-support system and communication gear so that the astronauts who are outside can talk to crew members in the spacecraft and to scientists and engineers at mission control on Earth. Space suits also provide oxygen and drinking water, which allows astronauts to work outside for several hours at a time to repair other spacecraft and satellites and to conduct scheduled experiments.

An astronaut's clothing must be versatile. Layering—wearing several thin layers of clothing that can be put on or removed as needed—is important for comfort and allows astronauts to adjust easily to changing conditions.

While clothing design and materials must meet certain standards, there is no reason for everyone to dress in exactly the same way. Each astronaut can choose colors that suit him or her. Astronauts can also choose such design features as the size and location of pockets. Wearing a cap from a favorite baseball team or a college T-shirt doesn't reduce the ability of crew members to complete their mission, but it does help each astronaut feel more like an individual.

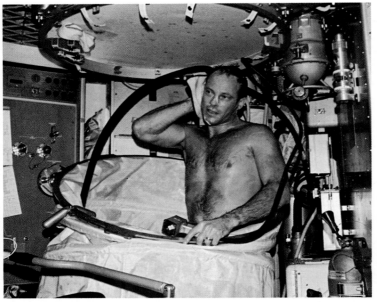

Astronaut Jack R. Lousma, above, takes a hot bath in the crew quarters of the orbital workshop of the Skylab space station in Earth orbit. To use the shower, he would pull up the shower curtain from the floor and attach it to the ceiling. The water comes through a push-button shower head attached to a flexible hose. Water is drawn off by a vacuum system. An engineer, right, demonstrates the hand-washing system in Skylab. Both hands are in the device, and he uses his right hand to start the water.

2
HYGIENE

On Earth, washing one's hands, taking a shower, shaving, and brushing one's teeth are usually done without much thought—almost automatically. In space, however, all these normal activities become much more difficult, and each astronaut must learn new ways to do them. Because there is no gravity, water won't pour or run down a drain.

Showers

Aboard the first small spacecraft, showers were out of the question, but the missions were so brief that personal hygiene wasn't a problem. With more recent flights on larger spacecraft, showers became a possibility. For example, on the Soviet *Mir* space station, cosmonauts took showers once every 10 days. On one mission they took them only once a month.

Taking a shower aboard a spacecraft isn't like taking one at home. On a spacecraft, water must be conserved, and preparing the equipment for showers takes a long time. On shower day, cosmonauts and astronauts spend much of their time on this activity.

A shower stall in a Soviet spacecraft consists of an elastic cylinder with caps at both ends. The caps contain devices for mixing and spraying hot and cold water, supplying hot air, and removing the used water. The stall has a waterproof zipper seal, which cosmonauts use to get in and out. Rubber slippers are fastened to the floor to keep the cosmonaut from floating upward.

Taking a shower is a complicated activity. First, a cosmonaut inserts a flexible hose into his or her mouth in order to breathe. Then the cosmonaut puts a clip on his or her nose to keep water out, because water doesn't flow downward in space. When the water is turned on, it comes out of holes at the top and bottom of the cylinder. Cosmonauts use a soap-filled cloth to wash, and then a plain cloth to rinse. After the water is turned off, a blast of air directs it out of the cylinder. Finally, the inside walls of the stall must be wiped dry.

Once all the crew members have taken their showers, the stall is collapsed like an accordion and raised up to the ceiling to be stored until the next shower day. Cosmonauts have occasionally complained about the difficulty of taking showers and have asked that future space stations have a permanent shower room that can be used at any time.

When the American space station *Skylab* was being designed, a contractor came up with a plan for a shower that would have cost $3 million! The plan was canceled, and a simpler one like the Soviets' shower was adopted.

Washing

Because astronauts can only take showers once a week at most, it's important that they wash thoroughly and often. To conserve the limited supply of water, astronauts use towels made of antibacterial cloth saturated with a disinfectant solution. They wet the towels with water when they are ready to use them.

Teeth Cleaning

Astronauts and cosmonauts use electric toothbrushes and nonfoaming toothpaste from a thin, collapsible, metal tube. They are encouraged to massage their gums with a special tissue or cloth made of antibacterial material, which can be wrapped around a finger. Special chewing gum, used after every meal, also helps keep teeth clean.

Toilet Facilities

Because of weightlessness, toilets on spacecraft are equipped with seat belts, footholds, and handholds to keep an astronaut in place. Since there is no gravity to pull waste down into the toilet, crew members turn on a vacuum (air suction) device. The vacuum pulls urine through a tube into a storage tank. The air used for the suction passes through a filter to eliminate odor and bacteria. Feces (solid waste) is also pulled by airflow to a storage tank—a type of space-age outhouse—where it is dried to control odor. The waste is either returned to Earth when the spacecraft lands or is periodically ejected into space. A fan prevents odors from lingering in the cabin air.

Hair Washing

Astronauts fasten special towels to a massage brush to wash their hair. This method is somewhat similar to the type of shampoo that they might get on Earth.

Shaving

While on a mission, crew members shave with a normal electric razor with rotating heads. In space, however, the razor also requires a special nozzle to vacuum the tiny shaved bits of whisker. Without the vacuum, these tiny bits of hair would fly around the cabin and

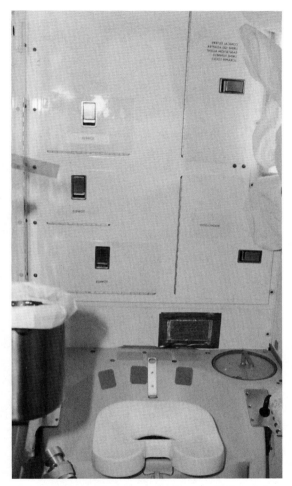

A toilet aboard the space shuttle Columbia

could work their way into small openings in vital equipment. That could cause the equipment to break down or malfunction.

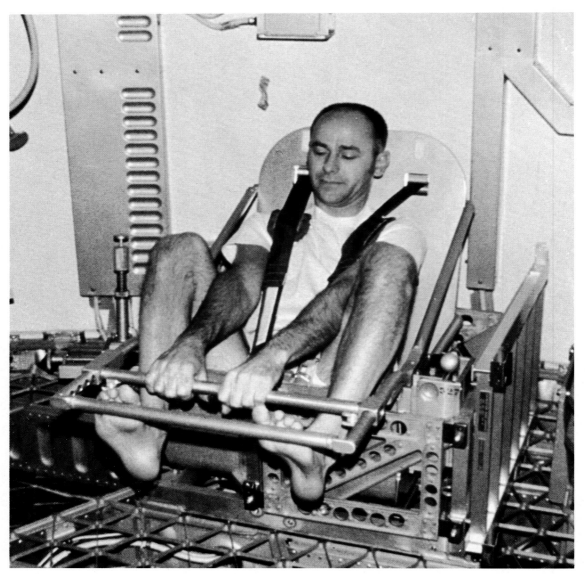

Aboard Skylab 3, *astronaut Alan L. Bean uses the body mass measurement experiment equipment.*

3
STAYING HEALTHY

On Earth everyone needs enough exercise, leisure time, and rest to stay in good health. The same is true in space, although living conditions there are different from those on Earth.

Exercise

Some activities considered normal exercise on Earth—such as walking, lifting heavy objects, and climbing stairs—require almost no effort in zero gravity. Space crews need other ways to keep their muscles and cardiovascular systems (heart and lungs) healthy. In addition to keeping astronauts healthy, exercise makes them feel better, and a good mental attitude leads to better work performance.

Because of the lack of gravity in space, some muscles are worked harder than usual as astronauts simply try to remain in place and do normal tasks, such as tightening bolts or unscrewing lids. Stationary bicycles, treadmills, and rowing machines—all fastened to the floor—can be found aboard a large space station. Equipment that involves lifting weights does not work in zero gravity, so springs or strips of elastic, which astronauts can pull against, must replace weights.

As is the case with many people, getting astronauts to exercise regularly can be a problem, especially if they are unusually busy doing experiments and running the spacecraft. It is important for the spacecraft commander to make sure everyone exercises long enough and hard enough. Body sensors attached to an exercising astronaut record vital signs that reveal how effective the exercise has been.

Astronauts need to exercise at least two hours a day, which is a longer period of time than some people would like. One former Soviet cosmonaut who flew on two six-month missions put it this way: "I hate our exercises. Loved it on Earth, but here, each time, I have to force myself. Boring and monotonous, and heavy work. But you realize you need it to keep in shape, so you grin and bear it."

It might be more interesting for the astronauts if several of them were able to exercise together—playing sports designed for space—but all the activity could shake the spacecraft and disturb delicate scientific experiments. However, two *Skylab* crewmen were once able to race around the cylindrical walls of the craft in a comical "*Skylab* 500," but that was an exception.

Leisure

Everyone needs time alone to relax and to do things that are personal and not necessarily productive—time to be an individual.

On the first brief space flights in tiny cabins, no one thought about leisure activity because there was no leisure time to be filled. The pioneers who explored the American West didn't need to take vacations in order to visit interesting places, because that was what they did all the time. So it was for the first astronauts who were doing things that no humans before them had ever done. They were hundreds of miles above the Earth, orbiting at 18,000 miles (28,800 km) per hour. The experience of being in space provided enough excitement and adventure for anyone.

But as space flights became more common—almost routine—just being "up there" was no longer enough to keep an astronaut from occasionally getting bored. Serious thought had to be given to recreation to break up the monotony and prevent boredom. Those whose interests included activities like playing football, going fishing, or flying model airplanes were out of luck. So were golfers and dancers. There are only so many things that can be done in the cabin of a spacecraft that has

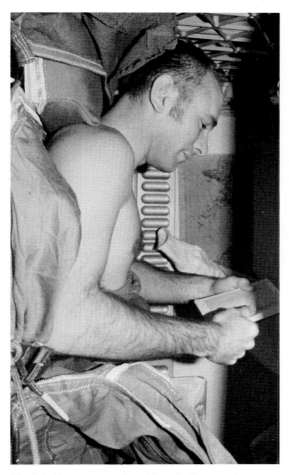

Skylab 3 *commander Alan L. Bean relaxes by reading a book in his sleep restraint.*

vented, either by mission planners or by the astronauts themselves.

Just about everyone likes some kind of music, and with the convenience of tapes and laser discs, it is easy to arrange for each astronaut to have a small music library. With earphones, one person's favorite music won't annoy other crew members. New tapes can be sent up in a supply craft or be transmitted by radio for copying aboard the spacecraft. Some crew members can even perform their own music on instruments carried on the spacecraft.

One of the most popular pastimes for space crews is reading, and books take up little room. Astronauts can trade books if they have similar tastes, and reading does not cause conflicts. Having someone next to you reading a book you don't like isn't the same as being forced to listen to someone else's favorite music—which you might not like.

Transmitting live, televised sports events to space vehicles lies somewhere in the future because live games tend to hold the attention of the viewer. Watching a live game could distract an astronaut from performing vital tasks. But written descriptions and still photos can be sent to the space shuttle,

no gravity. But with a little imagination, many familiar leisure activities can be transferred to space. Perhaps some completely new ones can be in-

The astronaut in this space station mock-up stands at the door of her personal compartment, which is equipped with a TV, headphones, a computer, and storage space.

and tapes of sports events, concerts, plays, and anything else the crew wants can be easily obtained. Such activities as playing chess and checkers are also possible and provide the astronauts with a complete change from their everyday duties.

Since the first cosmonaut went into space more than 30 years ago, sightseeing has been a favorite leisure-time activity aboard spacecraft. The view from hundreds of miles above the Earth is absolutely spectacular and constantly changes. A special viewing window was built into *Skylab*—over the objections of engineers, who didn't think it was a good idea. The crew made regular use of it, however, and never seemed to tire of looking down at their home planet.

With a two-way television hookup, it is possible for space crews to have private conversations with their families.

This extraordinary photograph of Earth was taken from the Apollo 17 *spacecraft. The view extends from the Mediterranean Sea to Antarctica's polar ice cap. Note the heavy cloud cover in the Southern Hemisphere. Almost the entire coastline of the continent of Africa is visible. The Arabian Peninsula can be seen at the northeastern edge of Africa. The large island off the southeastern coast of Africa is the Malagasy Republic. The Asian mainland is on the horizon toward the northeast.*

Rest Breaks

People who work hard or study hard need an occasional break to get up and stretch or have a soft drink or a cup of coffee. This is as true in space as it is on Earth. Long periods of concentration reduce efficiency and increase irritability. Even the most interesting work can sometimes become boring without a break.

In some jobs there is a rest break every few hours so employees feel and work better. Experience with crewed space flight has shown that astronauts need a short break at the first signs of fatigue. The break can be passive (doing absolutely nothing for a few minutes) or active (exercising or doing some other activity). What is important is having a change of pace.

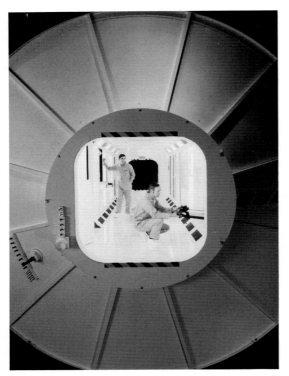

This full-scale mock-up of space station Freedom *shows the habitation module, where crews will spend their off-duty hours.*

This helps reduce homesickness, which can be a serious problem on a long flight during which crew members are cut off from their normal lives. Total separation for a long period of time can place serious strain on a marriage and other relationships, but frequent phone conversations can help.

Sleep

When you're really tired, it feels wonderful to flop down on a bed and sleep. In space, however, there is no "down." An astronaut sleeps in a standing position in a sleeping bag attached to the wall of a sleeping compartment. In the space station *Freedom*, elastic bands keep the sleeper in place and

Staying Healthy 37

The amount of sleep an astronaut gets is an individual matter, but most require about eight or nine hours a day. If there is extra work to do, sleep may have to be postponed until the next day, but a tired astronaut does not work well. At times it may be difficult for an astronaut to fall asleep. Occasionally he or she may have to take a sleeping pill, but astronauts most often can read a book to relax the mind and fall asleep.

A long-term lack of sleep causes serious problems. The more complicated and precise the work an astronaut has to do, the more important it is to be well rested. Once astronauts get used to sleeping in space, they dream regularly, and many of their dreams are about Earth.

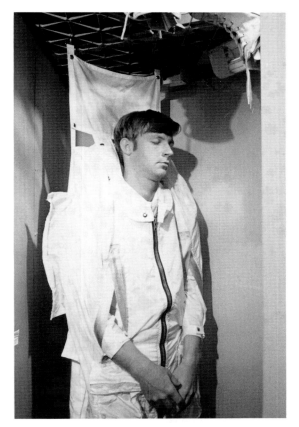

A technician demonstrates the sleep restraint system for use in space. Sleeping upright would be uncomfortable on Earth, but in zero gravity, there is no up or down. Note the sneakers in the top, right-hand corner.

give the feeling of lying horizontally on a bed because the bands press the sleeper back in the same way that gravity presses down.

Life aboard a spacecraft provides very little privacy. In this space station mock-up, one crew member sleeps in her private compartment while another works at his computer terminal. Behind the sleeping crew member is her personal effects bag, where she stores her belongings.

4
GETTING ALONG TOGETHER

Most people know that it's important to get along with family, classmates, and co-workers. In a cramped spacecraft, it is even more important to maintain friendly relations with those around you. Even if two astronauts develop a bad relationship, there is no escape from each other. Crew members must learn how to settle their differences calmly so that the mission can be completed.

Privacy

For many people, one of the most difficult things to get used to in a new environment is the lack of privacy. Whether living in a college dormitory, an army barrack, or a summer camp, being constantly with other people can be difficult. In space, too, privacy is important. In *Gemini*, the first American two-person spacecraft, privacy was impossible because quarters were so cramped. Astronaut Mike Collins said that the spacecraft was like the front seat of a Volkswagen Beetle. But as spacecraft became larger, private quarters for each crew member became possible.

In the American space station *Freedom*, each astronaut will have a soundproof compartment about seven feet by five feet by four feet. Each compartment will have a sleeping bag, a television with a video tape recorder, a stereo, a telephone, and a personal computer.

The longer a mission continues, the more important it is for each astronaut to have some private area to get away from other crew members, from the equipment and experiments, and even from the support crew communicating from Earth. Although it is vital to monitor the astronauts' actions at all times via radio and TV, this surveillance can make them feel like animals in a zoo. Given the conditions of space travel, astronauts also need some amount of privacy.

Who's in Charge?

Cosmonauts and astronauts are carefully selected from hundreds of applicants. They are then trained to fly the spacecraft and conduct important experiments. But to the crew in space, it often seems that the support crew on Earth makes all the big decisions. The careful planning required by any space mission is done mostly by scientists and engineers who will never go into space.

Sometimes the space crew wonders why they aren't allowed to think for themselves. This conflict sometimes leads to arguments among people who are all considered the very best in their fields. Who *does* know best, and who should make the final decisions?

Another debate revolves around the question of whether humans or machines work better in space. Most spacecraft are uncrewed. They are designed to do the same thing over and over for weeks, months, or years. Machines can do many simple jobs that would be boring to people. Other jobs, however, require such human qualities as creativity, originality, and the ability to think quickly and make decisions when unexpected events occur.

Humans, for example, can invent new experiments when the planned ones don't work or when the results of the planned experiments suggest a new approach. Humans can also correct mistakes that couldn't be predicted by planners who only simulated the mission in the laboratory.

Scheduling

On Earth most people follow a fairly regular schedule. They get up, get dressed, go to work or to school, come home, relax, eat dinner, and so forth. In space, schedules are needed to accomplish the work that needs to be done and to make crew members feel comfortable and "at home." If each astronaut independently decided when to do what, chaos would result. The steps of an experiment might be done out of order, and some jobs probably wouldn't be done at all.

> The schedule for a typical day on a *Salyut* mission is listed below:
>
> | Time | Activity |
> |---|---|
> | 8:00 – 9:40 | wake up |
> | 8:00 – 8:20 | test space station |
> | 8:20 – 9:00 | morning toilet (washing up, etc.) |
> | 9:00 – 9:40 | breakfast |
> | 9:40 –12:00 | work |
> | 12:00 – 1:00 | physical exercise |
> | 1:00 – 1:20 | free time |
> | 1:20 – 2:20 | work |
> | 2:20 – 3:00 | lunch |
> | 3:00 – 4:00 | free time |
> | 4:00 – 5:30 | work |
> | 5:30 – 6:00 | tea (a small meal) |
> | 6:00 – 7:00 | work |
> | 7:00 – 8:00 | physical exercise |
> | 8:00 – 8:30 | dinner |
> | 18:30 –11:00 | free time |
> | 11:00 – 8:00 | sleep |

Within this schedule, however, individual cosmonauts made minor changes to fit their own personalities. For example, not everyone needs nine hours of sleep every night. Some people need more than two hours of exercise on certain days, especially if they were too busy to exercise the day before. But major changes can create serious problems. To complete the important work of the mission according to schedule, astronauts must be awake and alert and perform specific jobs at the proper times. If astronauts did not coordinate their schedules, the experiments that had been planned would not get done.

Sex Roles

No women were aboard the first crewed space missions. Some people feared that including women in a crew might lead to serious problems. In 1963 Valentina Tereshkova became the first woman to attempt a solo flight. She orbited the Earth for three days, and then ejected herself from her capsule at a height of 4.2 miles (7 km).

A woman did not go into space again for almost 20 years. Veteran aerobatic and test pilot Svetlana Savitskaya was a

As one of their many tasks, astronauts studied solar astronomy aboard Skylab.

Mission control on Earth closely monitors and directs spacecraft crews. Sometimes crew members in space wonder who is actually in charge.

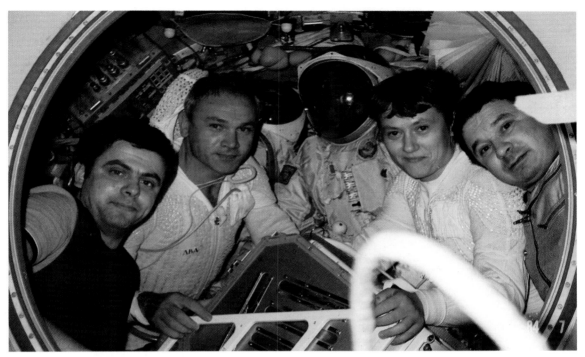

Svetlana Savitskaya, second from right, with the crew of the Salyut 7 *space station. Savitskaya said, "Women go into space because they measure up to the job."*

visiting cosmonaut aboard the space station *Salyut 7* on its seven-month mission in 1982 and on its eight-month mission in 1984. The chief of the Cosmonaut Training Center said:

> We have noticed that in training... the whole work atmosphere and the mood of the crew of men and women [is] better than in a men-only one. Somehow the women elevate relationships in a small team, and this helps stimulate its capacity for work. When Savitskaya was aboard *Salyut*, the five male crew members acted differently than usual.

This attitude did not please her. In a news report following her return to Earth, Savitskaya "appeared slightly irritated by frequent references to the pleasant atmosphere a woman brings to a space station." She said, "We do not go into space to improve the mood of the crew. Women go into space because they measure up to the job."

Group Dynamics

Advance preparation and planning help crew members avoid annoying each other and diminishing the success of the mission. Space crews must get to know one another's personalities and capabilities, and they must learn to understand one another by a word or a glance. Teamwork is at least as important in space as it is anywhere in life. The crew must get through stressful periods without anyone "blowing up." And crew members must be able to endure periods of boredom, isolation, and confinement.

The crew must be carefully tested before and during their training for the mission. Putting crew members under pressure while they're training on Earth will give them and the flight planners a sense of how they will get along in space. After a flight has been completed, each crew member must give full details of how she or he got along with the other crew members.

Astronauts do not need to have similar personalities in order to have a smooth mission. Sometimes people with very different personalities get along well in close quarters—even when they're under considerable pressure. Knowing a lot about each other in advance makes it possible to plan for potentially tense situations.

Once a crew is in space, the members have to deal with problems as they arise. The situation is much like setting off on a long cross-country automobile trip with people you don't know very well. If everyone realizes how unpleasant the trip will become if there are big disagreements, then they'll usually try to maintain peace during the journey. Just one person constantly demanding his or her own way can make a journey tense and unpleasant. On a short space flight, consideration of others' needs is fairly easy; on a six-month mission, maintaining good relationships can be more difficult.

On the seven-month mission of *Salyut 7* in 1982, the crew reportedly had some difficulties getting along together. Group training was short, so many personality conflicts had to be worked out in space. Friction among

Getting Along Together 45

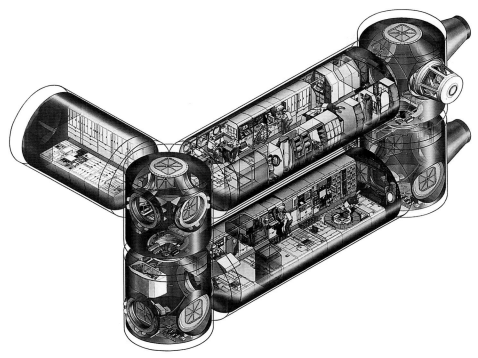

This cutaway of planned space station modules shows the living area at top center, the laboratory below, logistics storage at left, and interconnecting nodes at the ends of the main modules.

crew members sometimes interfered with the smooth operation of the mission. Prior to *Salyut 7*'s eight-month mission in 1984, an improved selection process for crew members led to better working relations among the cosmonauts.

Relations among the crew aren't the only important ones. Crew members in space and people in mission control on Earth must also cooperate with and support each other. There have been instances in both the United States and the former Soviet programs when important information was not transmitted from space to ground for personal reasons. For example, a case of severe motion sickness was not communicated to avoid embarrassment. Such behavior did not contribute to the success of those missions.

An engineer and an electronic technician check the installation of equipment racks and environmental control systems in a laboratory module simulator (LMS) at the NASA Marshall Space Flight Center in Huntsville, Alabama.

5
DESIGNING A LIVABLE SPACECRAFT

Most people are used to having enough space to live in—to work, to study, to sleep, and to play. Even when we ride in cars during long trips, we make frequent stops to stretch our legs or toss a ball around.

Those conditions aren't possible in space, however. Astronauts are locked into a vehicle that has been designed to carry as many people and as much equipment as possible. Because the cost of launching a spacecraft depends on its size and weight, there can't be much "wasted" space. To make a spacecraft affordable, everything must be as streamlined as possible. People, however, need room in which to relax and unwind. If astronauts bump into each other every time they move, they might begin to feel like caged animals, and they will probably not be in a very good mood. A spacecraft crew that is in a bad mood won't be able to do its job well. The newer spacecraft are designed with the comfort of the crew in mind, as well as the efficiency of the vehicle.

Lighting the Cabin

Many factors besides the amount of space contribute to the general well-being of people. Too much light, for example, can hurt your eyes. Too little light can make reading and working difficult. Unnatural light can make people very uncomfortable, as can light of the wrong color—even if you don't realize it. Different activities, such as eating, working, playing, and sleeping, require different amounts of light. In addition, each individual, including those in a spacecraft crew, has his or her own special preferences for lighting.

Different *types* of lights—floodlights, spotlights, and pinpoint lights, for example—are needed for different settings and activities. Light must also be balanced with the color and reflectivity of the walls of a room to make the effect as pleasing as possible. Proper lighting is especially important in a spacecraft.

The former Soviets found, for instance, that reading with yellow or green light strains the eyes less than reading with red light. They also discovered that, as a mission wears on, their cosmonauts wanted more and more light. The best lighting is that which comes from above, like natural light from the Sun.

Windows

Anyone who has ever worked or spent much time in a room that has no windows knows how annoying it can be. You are unable to tell if it is light or dark, or rainy or sunny outside, and you can feel cooped up. The same holds true for the crew of a spacecraft.

The windows on early spacecraft were easily scratched or damaged on the outside by micrometeorites. On the inside, particles and dust often coated the windows and made it difficult to see through them. Improvements have been made, however, and now windows retain their clarity much longer. Special chemical coatings on windows also reduce the formation of bacteria inside the spacecraft.

Looking out a window at the Earth below is one of the most enjoyable activities for astronauts in space. It is not only an educational pastime, but a very relaxing one, too.

Decor

The inside of any spacecraft is crammed with scientific equipment and instruments designed to be useful rather than attractive. In the new, larger craft, however, more and more interior areas are being treated to make them look pleasing and more like home. The scientific use of color, in particular, has a strong impact on astronauts' morale. Color can improve the way astronauts feel about their jobs, keep them more alert, and reduce the chances of

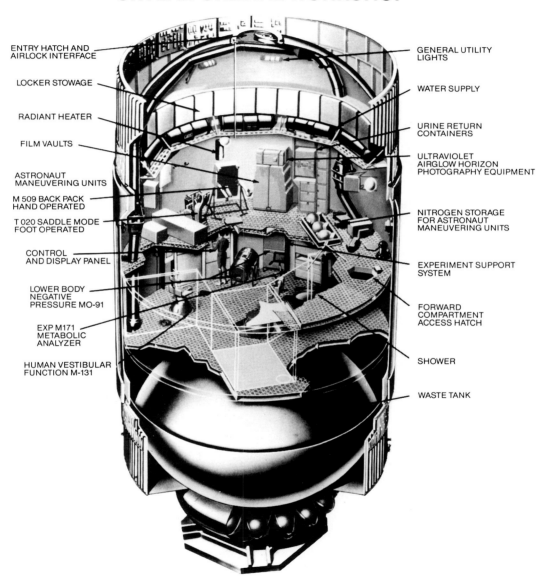

50 LIVING IN SPACE

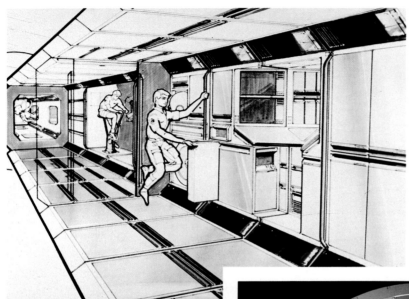

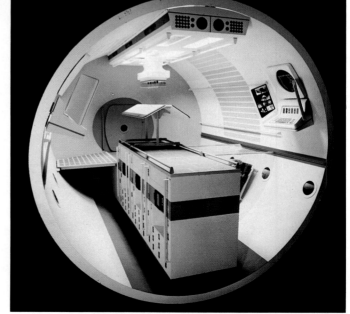

An artist's concept of the living area of a space station is shown above. In the foreground, a crew member prepares a meal in the kitchen, while behind him another astronaut gets a workout on an exercise bike. In the background, a third crew member enters a sleep compartment. To the right is a mock-up of a space station module. Note the use of color in the module's design.

behavioral problems resulting from long-term confinement.

The inside of one *Salyut* space station was painted in soft pastel colors to give it a homey feel. The next *Salyut* station had a completely different color scheme. In the working compartment, one wall was painted apple green and the opposite wall beige, accented with bright orange, blue, and white. The ceiling was painted white. This color scheme provided a bright, pleasant working environment.

Warm tones (red, orange, yellow) are thought to stimulate the nervous system, while cool tones (blue, green, violet) are thought to be soothing. Because using one solid color can make a large area look monotonous, walls and surfaces are often painted with different shades and tones of a single color. Small areas might be painted in bold colors. Wall hangings, paintings, and drawings can also help make an area look more interesting and less mechanical.

Research shows that warm, sunny tones are good colors for the living quarters of a crewed craft. Warm, relaxing colors with a contrasting trim are good for rest and recreation areas. In sleeping areas, cool, muted colors create a feeling of coziness and the impression of increased space.

Cabin Odor

When the air in your room or office smells bad or gets stuffy, you can usually open a window or door. But opening windows in a spacecraft is impossible. Even a slightly bad smell can have a powerful effect on the morale of the crew if it isn't dealt with quickly. Smells can come from food and from the astronauts themselves, especially after strenuous exercise.

Sometimes filters in the air-conditioning system can help get rid of odor problems. Careful selection of food also helps, but some of the tastiest foods often produce strong odors.

Washing is the best solution for body odor, but astronauts can't take daily showers because there isn't enough water on board. Instead, they have to clean themselves regularly with special, moist washcloths that have been treated to eliminate bacteria. These cloths are then stowed away in sealed containers so their smells won't spread throughout the cabin.

Regeneration of the Cabin Air

In a spacecraft, the air itself can pose a major problem. For example, if the chemical composition of the air changes, the impact on the crew can be serious. When people breathe, they inhale oxygen and exhale carbon dioxide. Each astronaut breathes in about seven gallons (23 l) of oxygen and breathes out about five gallons (19 l) of carbon dioxide each day. Over a period of time, the amount of oxygen in the spacecraft drops and is gradually replaced by carbon dioxide. On Earth this process is balanced by photosynthesis, a natural phenomenon in which plants take in carbon dioxide and give off oxygen.

In spacecraft, filters are used to control the chemical composition of the air. The equipment aboard a spacecraft often produces chemicals that affect the air. Astronauts must change canisters of lithium hydroxide and carbon frequently. The canisters cleanse carbon dioxide and odors from the air.

Foul air can make astronauts sick and can irritate their eyes, thus reducing their ability to function. When this happens, the air must be replaced by either a crewed or uncrewed service vehicle. In other words, fresh air has to be delivered to the spacecraft from Earth. During some of their long missions, the former Soviets have used this procedure.

Sound and Vibration

The combination of sound and vibration—especially if it is continuous and unchanging—can become very irritating to the crew. Many sounds cannot be helped. Indeed, changes in sound quality and level can be valuable by warning that some piece of equipment is not working correctly and should be checked. However, too much sound, no matter what kind, can produce irritability, sleep disorders, headaches, fatigue, and even temporary or permanent loss of hearing.

Music can help vary the sound level and add a pleasing element to what otherwise might be considered annoying. In addition, the careful use of insulating and sound-absorbing materials can lower the level of sound. Some equipment, and even whole spacecraft, can be designed to minimize noise and vibration. In situations where that is

Astronaut Dr. Owen Garriott eats a meal while in orbit aboard Skylab 3.

too difficult or expensive, crew members may have to wear earplugs or ear protectors at least part of the time.

Temperature Control

Temperatures vary greatly around an orbiting spacecraft as it moves quickly from brilliant daylight to total darkness. A complete orbit around the Earth takes only 90 minutes, and the crew sees 16 sunrises and 16 sunsets during the course of a 24-hour day.

Temperatures outside the spacecraft range from 264°F in full sunlight to -236°F in Earth's shadow (128°C to -148°C). Inside the spacecraft, heat is given off by much of the equipment and also by each crew member. Radiators and air blowers keep the cabin temperature between 65° and 80°F (18° and 27°C) and the humidity between 20 and 80 percent.

On Skylab 4, astronauts Gerald Carr, left, and William Pogue, right, stuff bags of trash into the waste disposal tank.

6
SAFETY AND MAINTENANCE

In space, making even a small error can place the crew and the entire mission in jeopardy. Safety training in space is extremely important. A momentary lapse could mean the difference between success and disaster.

For example, a careless docking maneuver could damage both the orbiting craft and the smaller space vehicle. Such a mishap would not only end the linkup attempt and leave the crew short of supplies, but would also endanger the lives of the crew.

Crew members don't have to wear special clothing when they're working inside the spacecraft. For working in the cargo bay or taking space walks, however, astronauts need space suits with portable life-support systems built into them. Even a pinhole in a pressurized space suit could cause air to escape, and the astronaut would quickly suffocate.

The larger and more complicated a spacecraft is, the greater the chances of a possible disaster. In space station *Freedom*, plans call for special areas where the crew could go in case of an accident such as a fire or an explosion, or if the space station were hit by a micrometeorite or a piece of space debris. A shuttle craft to be used for rapid evacuation may be permanently docked at the station.

Maintenance

The design and construction of a sophisticated space station requires equipment that operates well without needing frequent repairs. To complete its complicated experimental work in the scheduled amount of time, the crew of a space station can't spend much time maintaining or repairing the spacecraft, nor can they call a repair person when something stops working. The crew has to locate the problem and then fix it. When repairs are necessary, the faulty parts must be easy to locate and fix. Tools and spare parts must be labeled carefully and stored where they can be found quickly when needed.

Because the lack of gravity makes tools more difficult to use, repairs can take longer in space than they would on Earth. Astronauts must also save, label, and store the broken or malfunctioning parts so that engineers can analyze them—figure out why they failed and improve them for use on the next spacecraft.

Space crews see themselves as scientists, engineers, pilots, and explorers —not as repairers. The less time spent fixing a spacecraft, the happier the crew will be. However, engineers often create complicated devices, such as specialized tools that perform only one specific job. Finding and using such tools can annoy astronauts, so they often work out their own simpler solutions. On one of the *Salyut* missions, for example, a pair of ordinary scissors became the most important tool on board, because it could be used for so many different jobs. On another mission, the specially trained crew that was supposed to join the *Salyut* and install solar panels was unable to make the trip. Cosmonauts who hadn't been trained for this complicated job installed the panels with the aid of instructions relayed from the ground by radio. So far, astronauts have been extremely creative at "making something out of nothing" and inventing new ways of doing things—ways never dreamed of before the launch.

Housekeeping

As on Earth, one of the least popular duties aboard a spacecraft is housekeeping. Astronauts use a vacuum cleaner and special dust cloths to keep everything as clean as possible. But no matter how carefully the crew tries, food crumbs, scraps of paper, and other tiny objects sometimes fly around the cabin and find their way into delicate machinery and other out-of-the-way places. If the particles aren't cleaned out immediately, equipment can short-circuit or jam, forcing the crew to spend extra time on maintenance.

In the words of Cosmonaut Valentin Vital'yevich Lebedev, "It might seem like a trifling matter to put things in their place and secure them properly. But this kind of thing is no longer trifling under conditions of weightlessness and can cause pure wasted time." Tasks that on Earth are as simple as sewing a torn strap can become a major problem in orbit because of the absence of gravity. Practice and a steady hand are needed.

Emergencies

Because aid from outside the spacecraft would take at least several days to arrive, astronauts must be prepared to deal with emergencies and minor problems that could turn into emergencies. Sometimes the problems encountered are psychological ones.

For example, a *Soyuz* shuttle that was to carry supplies to the *Salyut 6* space station developed engine trouble and had to return to Earth. The crew aboard the space station knew that they would later use another *Soyuz* shuttle to return to Earth. Because both spacecraft had the same type of engine, the crew felt the distinct possibility that their *Soyuz*, too, could encounter problems.

At first, the *Salyut* crew began to worry about the engine, which could not be tested before use. Had they become overwhelmed with anxiety, they could not have performed their tasks and would not have been thinking clearly. According to Cosmonaut Valeriy Viktrovich Ryumin, "...two possibilities opened up to us at this point in time: Let ourselves go...and fall apart; or grin and bear it, and don't give it another thought. We chose the second, as a means of psychological survival. And to make sure it worked, fell into work, leaving ourselves no time for tormenting doubts." The engine worked well on the flight home, and the cosmonauts finished their work in space without being devoured by fears of what might happen.

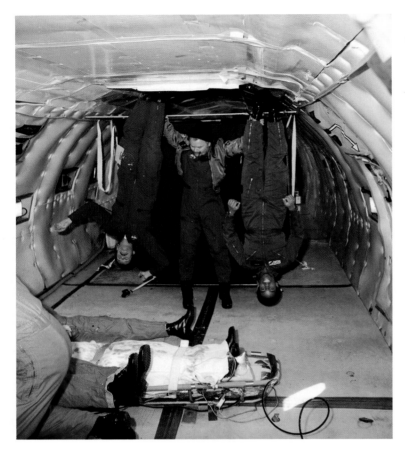

Two astronaut candidates get a unique view of their environment while being tested during a zero gravity flight. In the lower left-hand corner another trainee conducts medical studies and tests for motion sickness.

Medical Emergencies

The longer a mission in space extends, the greater the chances of a serious illness or accident occurring among the crew members. Although well-trained astronauts can deal with minor medical problems, more serious problems require a doctor. Until space crews include trained medical people, having extensive medical information and advice available at short notice must suffice. So far, no major medical problems have arisen in space.

Safety and Maintenance 59

This computer-generated image shows how NASA's space station may look when it is deployed in the 1990s. A space shuttle is shown approaching the station. Boeing Aerospace and Electronics is designing and building the station's living quarters and work areas, as well as the logistics modules that will be used for storage and carrying supplies to and from the station.

SUMMARY

In only 30 years, astronauts have progressed from being highly trained test pilots, who had little control over their small, crude craft, to being scientific specialists who concentrate on performing complex experiments in sophisticated, orbiting laboratories.

As space flight becomes a more common part of life—rather than a fantastic adventure—the quality of life inside a spacecraft will continue to improve. No longer do astronauts spend all their waking hours flying the craft and operating its equipment. They are human beings whose various needs must be met while they are temporarily living in space.

If astronauts are to do their jobs effectively, their personal lives, problems, and individual needs must be dealt with as seriously as their experiments. In the future, space flight is likely to become one of the most exhilarating experiences known to human beings.

RESOURCES TO CONTACT

American Astronautical
  Society, Inc.
6352 Rolling Mill Place
Suite 102
Springfield, VA 22152

American Institute of
  Aeronautics and Astronautics
370 L'Enfant Promenade, S.W.
Washington, DC 20024

Boeing Aerospace and
  Electronics
Public Relations
P.O. Box 3999, M/S 85-19
Seattle, WA 98124

Center for Aerospace Sciences
University of North Dakota
P.O. Box 8216,
  University Station
Grand Forks, ND 58202

The Commission on Science
  and Technology
U.S. House of Representatives
2321 Rayburn House Office
  Building
Washington, DC 20515

Eyes on Earth
146 Entrada Drive
Santa Monica, CA 90402

International Astronautical
  Federation
3-5 rue Mario Nikis
Paris Cedex 15 F-75738, France

John F. Kennedy Space Center
  and Library
SAN 302-9905
Kennedy Space Center, FL
  32899

Lunar and Planetary Institute
3303 NASA Road 1
Houston, TX 77058

National Aeronautics and
  Space Administration
400 Maryland Avenue, S.W.
Washington, DC 20546

National Air and Space Museum
Smithsonian Institution
7th Street and Independence
  Avenue, S.W.
Washington, DC 20560

National Space Club
655 15th Street, N.W., No. 300
Washington, DC 20005

National Space Society
West Wing, Suite 203
600 Maryland Avenue, S.W.
Washington, DC 20024

The Planetary Society
65 North Catalina Avenue
Pasadena, CA 91106

The Senate Subcommittee on
  Science, Technology, and
  Space
Hart Senate Office Building,
  Room 427
Washington, DC 20510

The Space Center
P.O. Box 533
Alamogordo, NM 88311-0533

Space Science Center
University of New Hampshire
DeMerritt Hall
Durham, NH 03824

U.S. Space Education
  Association
P.O. Box 1032
Weyburn, SK, Canada S4H 2L3

INDEX

Apollo, 10
Atkov, Oleg Yur'yevich, 13

Bolden, Charles, 17

clothing, 23-25
    for early space crews, 23
    personal preferences, 25
    for space walks, 25
    versatility of, 25
    worn in spacecraft, 24, 25
Collins, Mike, 39
Controlled Ecological Life Support System (CELSS), 23
cooking equipment, 19
Cosmonaut Training Center, 13, 43

decision making, 40
Discovery, 17

eating, 15-19
    during early space flights, 15
    and morale, 15
exercise in zero gravity, 31-32
    aboard *Skylab,* 32
    amount needed, 32
    equipment for, 31
    monitoring effect, 32
    need for, 31

food on space flights
    amounts of, 16-17
    cultural differences, 18
    growing, 19, 22-23
    individual preferences, 17-18
    preparation of, 19
    repetition of, 17, 18
    snacks, 18-19
    on former Soviet flights, 18
    storage of, 16
    supplements, 18
    testing, 15
    types of, 16
    value, 16-17
    waste, 19
Freedom 7, 8, 19
Freedom space station, 21, 37, 39, 55

Gagarin, Yuri, 7
Gemini, 39
Gemini program, 16
Glenn, John, 8
group dynamics. *See* space crews

hair washing, 29
Hawley, Steven, 17
homesickness, 36
Hubble Space Telescope, 17

Lebedev, Valentin Vital'yevich, 57
leisure activities, 32-34

McCandless, Bruce, 17
Mercury program, 8, 16
Mir space station, 27
mission control, 8, 25, 40

National Aeronautics and Space Administration (NASA), 12, 23

privacy on a spacecraft, 39-40

rest breaks, 36

Salyut mission, 41
Salyut 7, 13, 43, 44-45
Savitskaya, Svetlana, 41, 43-44
scheduling. *See* space crews

sex roles. *See* space crews
shaving, 29
showers aboard spacecraft, 27-28
   equipment for, 27
   the process of, 28
Shriver, Loren, 17
simulators, 8
Skylab, 12, 32, 34
sleep, 36-37
   amount needed, 37
   position during, 37
Soyuz, 16, 57
spacecraft
   air quality in, 52
   cabin odor, sources and control of, 51
   color, use of, 48, 51
   housekeeping, 56-57
   inside design, 47-53
   lighting, 47-48
   repairs aboard, 56
   sound, 52-53
   temperature control, 53
   vibration, 52
   windows, 48
space crews
   diet of, 15-19
   emergencies among, 57, 58
   group dynamics, 44-45
   safety measures, 55
   scheduling activities, 40-41
   sex roles among, 41, 43-44
   tension among, 8, 10, 44-45
   testing, 44
   workload, 8, 17
space station. *See Freedom, Mir,* and *Skylab*
space travel, early days, 7
   living conditions, 7
STS-31, 17, 18
Sullivan, Kathryn, 17, 18

teeth cleaning, 28
Tereshkova, Valentina, 41

toilet facilities, 28-29

Vostok 1, 7

washing, 28
weightlessness, 12-13

zero gravity, 13, 31

ACKNOWLEDGMENTS

All photographs in this book appear through the courtesy of the National Aeronautics and Space Administration (NASA), except for the following: pp. 21, 22, 34, 36, 38, 45, 46, 57, Boeing Aerospace and Electronics; p. 43, Tass from Sovfoto.
Front and back cover photos are courtesy of NASA.